I0901556

Extreme Weather

TSUNAMIS

Martha London

DiscoverRoo
An Imprint of Pop!
popbooksonline.com

abdobooks.com

Published by Pop!, a division of ABDO, PO Box 398166, Minneapolis, Minnesota 55439. Copyright © 2020 by POP, LLC. International copyrights reserved in all countries. No part of this book may be reproduced in any form without written permission from the publisher. Pop!™ is a trademark and logo of POP, LLC.

Printed in the United States of America, North Mankato, Minnesota.

052019
092019

THIS BOOK CONTAINS RECYCLED MATERIALS

Cover Photo: Dinodia Photos/Alamy

Interior Photos: Dinodia Photos/Alamy, 1; Alex Mustard/naturepl.com/Nature Picture Library/Alamy, 5; Kyodo News/AP Images, 6–7, 13, 30, 31; iStockphoto, 8, 9, 10–11 (background), 10–11 (boat), 15, 16, 21, 22–23, 27, 29 (wave), 29 (checkmarks), 29 (car), 29 (mountain), 29 (road), 29 (boat), 29 (television), 29 (bottom); Earth Observatory of Singapore, 14; Gemunu Amarasinghe/AP Images, 17; Shutterstock Images, 19, 20–21; Dadang Tri/Reuters/Newscom, 25; Shuji Kajiyama/AP Images, 26; Susan Watts/NY Daily News Archive/Getty Images, 28

Editor: Connor Stratton
Series Designer: Sophie Geister-Jones

Library of Congress Control Number: 2018965061

Publisher's Cataloging-in-Publication Data

Names: London, Martha, author.

Title: Tsunamis / by Martha London.

Description: Minneapolis, Minnesota : Pop!, 2020 | Series: Extreme weather | Includes online resources and index.

Identifiers: ISBN 9781532163975 (lib. bdg.) | ISBN 9781532165412 (ebook)

Subjects: LCSH: Tsunamis--Juvenile literature. | Tidal waves--Juvenile literature. | Severe storms--Juvenile literature. | Natural disasters--Juvenile literature.

Classification: DDC 551.4637--dc23

WELCOME TO DiscoverRoo!

Pop open this book and you'll find QR codes loaded with information, so you can learn even more!

Scan this code* and others like it while you read, or visit the website below to make this book pop!

popbooksonline.com/tsunamis

*Scanning QR codes requires a web-enabled smart device with a QR code reader app and a camera.

TABLE OF CONTENTS

CHAPTER 1

HOW TSUNAMIS FORM

A tsunami is a group of huge waves that hit land. Most tsunamis are caused by earthquakes. These earthquakes happen deep underwater. The ocean floor is made up of many **tectonic plates**.

WATCH A VIDEO HERE!

These huge plates move very slowly. Sometimes two plates run into each other. When that happens, those plates can cause an earthquake.

A diver swims in a gap between two of Earth's tectonic plates.

An ocean earthquake causes waves. At first, these waves are small and fast. But the ocean is shallower near the shore. The waves slow down as they

Tsunami waves often start less than a foot tall. Near shore, they can become more than 100 feet tall!

hit the shallow water. As these waves slow down, they get taller. When a tsunami finally strikes land, the waves can be massive.

Earthquakes cause nearly 75 percent of tsunamis. But tsunamis can have a few other causes. They can begin after a **landslide** under the ocean. A **volcanic eruption** can also cause a tsunami. These events also create waves.

DID YOU KNOW?

***Tsunami* means "harbor wave" in Japanese.**

Landslides can happen below water or on land.

Nearly 2,000 active volcanoes are currently on Earth.

FROM AN EARTHQUAKE TO A TSUNAMI

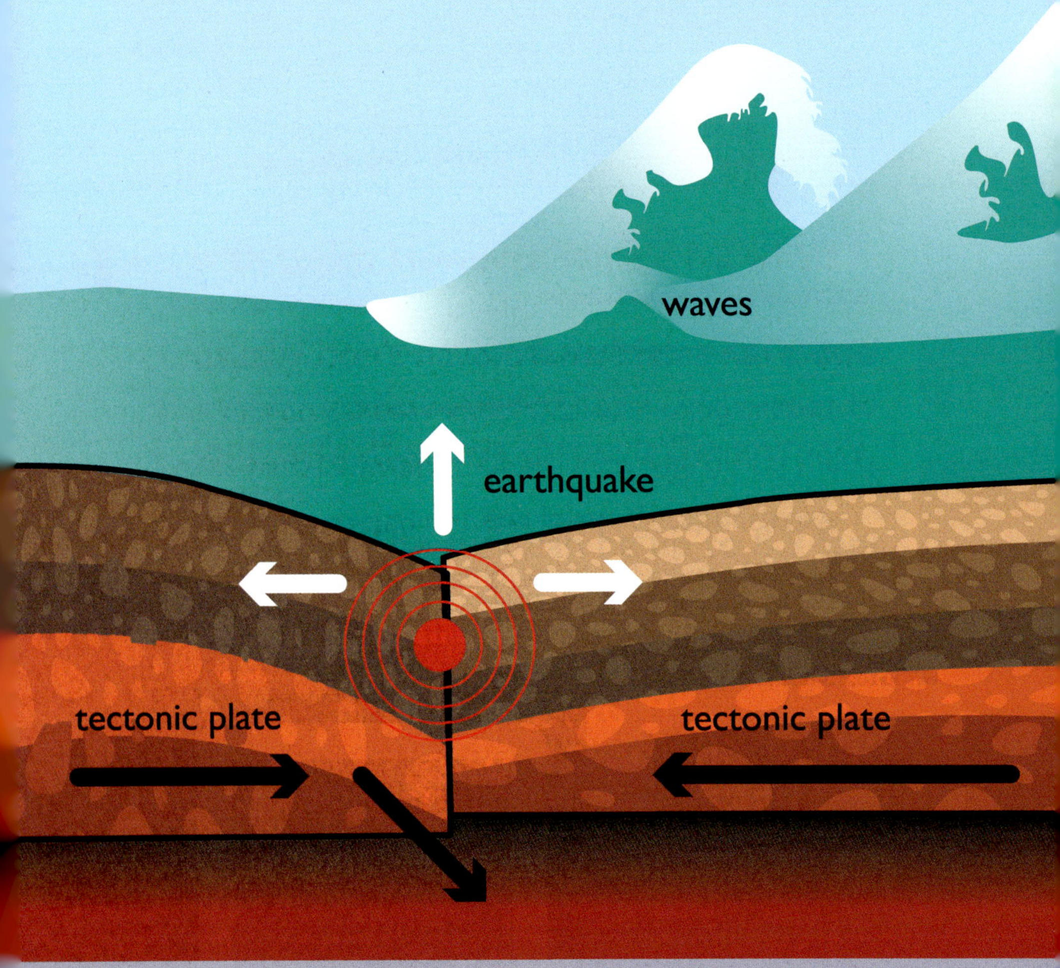

When two tectonic plates meet, one plate sometimes pushes the other under it. A strong earthquake often follows. This earthquake can cause a tsunami. The waves of the tsunami get larger near the shore.

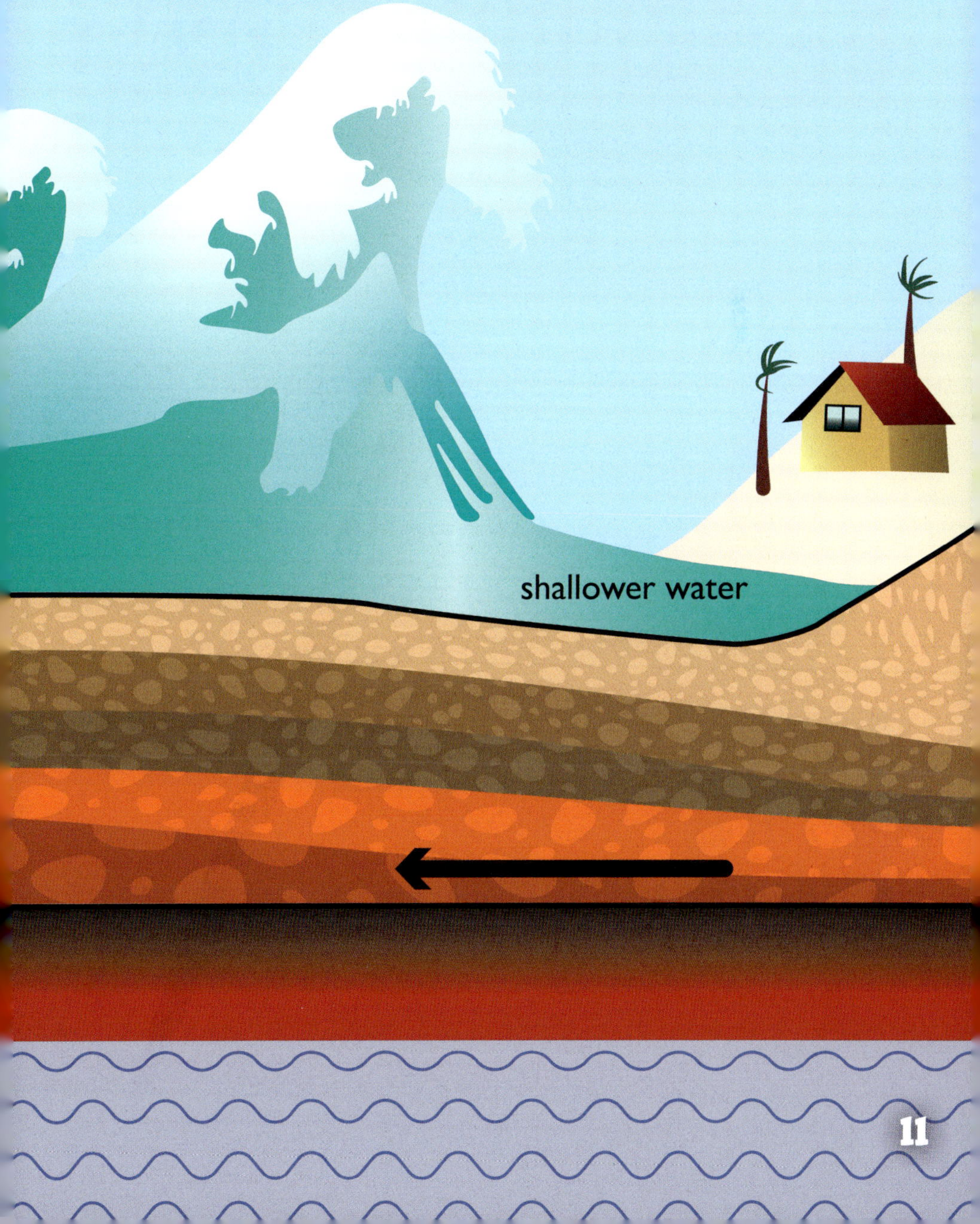

CHAPTER 2

WHERE TSUNAMIS FORM

Most tsunamis happen in the Pacific Ocean. These tsunamis hit places with lots of shoreline. For this reason, Japan, Alaska, and Chile are at high risk.

DID YOU KNOW?

Since 1900, one out of every five tsunamis has hit Japan.

In 2011, a powerful tsunami in the Pacific Ocean hit Japan.

LEARN MORE HERE!

Many **tectonic plates** lie below the Pacific Ocean. The Pacific Ocean also has many volcanoes. These volcanoes and plates stretch around much of the Pacific Ocean. This area is called the Ring of Fire. The Ring of Fire's volcanoes and

DID YOU KNOW?

Tsunami waves can travel as fast as a jet. They can cross the entire Pacific Ocean in less than a day!

Approximately 450 volcanoes are in the Ring of Fire.

earthquakes cause most of the tsunamis in the Pacific Ocean.

Tsunamis happen less often in other bodies of water. But tsunamis do occur in these areas. For example, a tsunami happened in the Indian Ocean in 2004. The tsunami caused horrible damage in Indonesia and several other countries.

TSUNAMIS IN THE CARRIBBEAN

Tsunamis are not common in the Caribbean Sea. But they can happen there. The area sometimes gets earthquakes and **volcanic eruptions**. In 1867, an earthquake caused a tsunami in the Caribbean. Some waves were as tall as a four-story building!

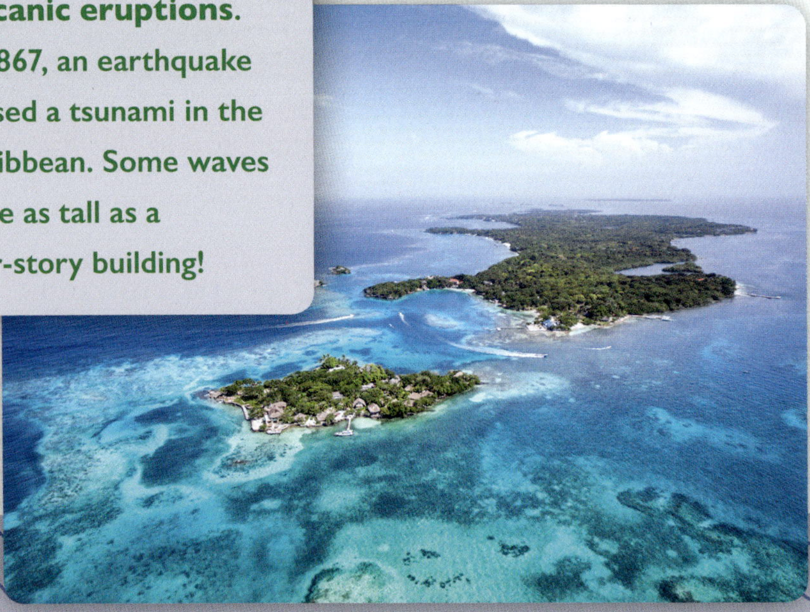

After the Indian Ocean tsunami in 2004, many homes in Sri Lanka flooded.

CHAPTER 3

DAMAGE

A tsunami can cause major damage. When it hits land, a tsunami can move at approximately 30 miles per hour. At that speed, water can destroy trees, cars, and even buildings. If water knocks over power lines, cities can lose power.

LEARN MORE HERE!

A tsunami's waves can carry boats and other large objects.

Staying at home during a tsunami may not be safe. A tsunami can lead to flooding. People can drown. Floods may release **sewage** into the water.

A tsunami's waves can flood and topple buildings.

This sewage can affect an area's drinking water. Unsafe drinking water causes disease to spread.

In addition, tsunami waves flow back to the ocean after they hit land. This water can sweep up people and objects. The water can carry those people and

Erosion took away soil from the roots of this tree.

objects out to sea. Tsunami water can also cause the soil to **erode**.

DID YOU KNOW?

In 1958, a tsunami in Alaska reached a record-setting height of 1,720 feet. Its waves carried away millions of trees.

CHAPTER 4

STAYING SAFE

Scientists study earthquakes in order to **predict** tsunamis. Many places send out tsunami warnings. These warnings allow people to prepare for tsunamis. People can then **evacuate** the area.

COMPLETE AN ACTIVITY HERE!

Scientists place devices in the ocean to tell them when ocean earthquakes occur. These devices help scientists predict tsunamis.

DID YOU KNOW?

One of the first signs of a tsunami can be a sudden drop in water level at the shore.

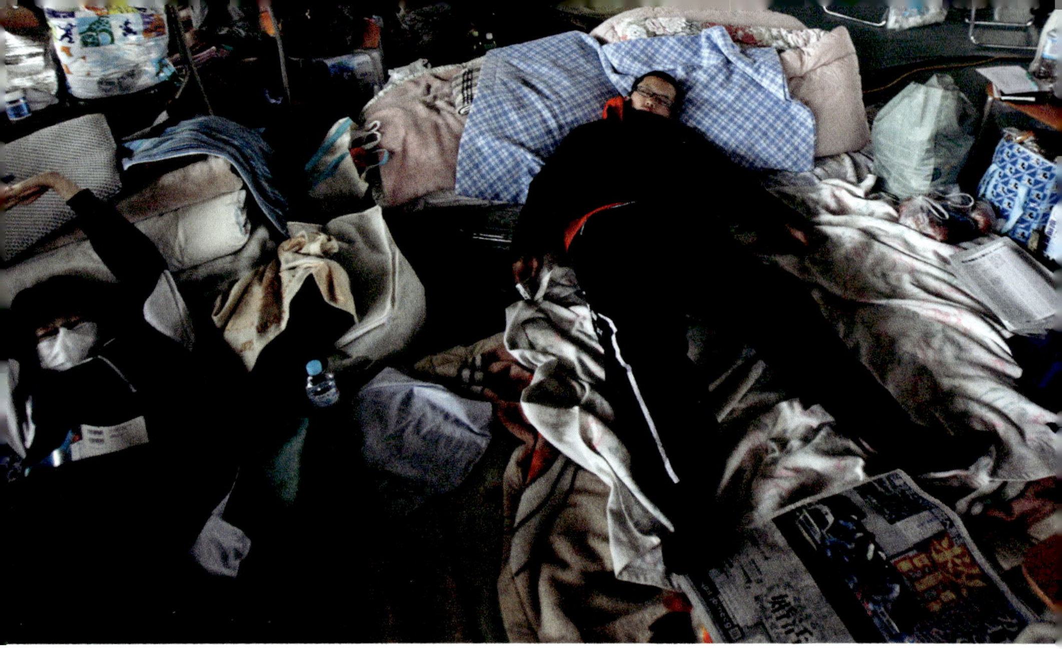

During a tsunami, people sometimes stay in shelters to stay safe.

Sometimes people cannot evacuate. Instead, they try to find higher ground. They also move away from the shore. Waves are less likely to hit these areas.

Some people may be on boats when a tsunami hits. These people travel farther out into the ocean. Tsunami waves are smaller away from shore.

Far out at sea, tsunami waves are usually very small. Sailors may not even notice a tsunami wave as it passes.

Electricity spreads easily through water. For this reason, water can be dangerous even if it is not directly touching power lines.

An area can still be unsafe after a tsunami. Floodwaters can remain for a long time. These waters can be dangerous. For example, destroyed power lines can be in the water. If that happens, water can shock people. People avoid flooded areas to stay safe.

Tsunami Safety Checklist

- ✓ Evacuate if necessary.
- ✓ Move to higher ground.
- ✓ Move as far inland as possible.
- ✓ If you are in a boat, go out to sea.
- ✓ Avoid flooded or moving water.
- ✓ Follow the news for weather updates.

MAKING CONNECTIONS

TEXT-TO-SELF

Are tsunamis common where you live? Why or why not?

TEXT-TO-TEXT

Have you read books about other kinds of natural disasters? How are they similar to tsunamis? How are they different?

TEXT-TO-WORLD

Tsunamis bring large waves onto land. Why does going to higher ground or away from shore help people stay safe?

GLOSSARY

erode – to slowly wear away.

evacuate – to leave a place in order to stay safe.

landslide – the sliding down of rock or earth from a mountain or hill.

predict – to say what will happen in the future.

sewage – human waste that is put into sewers.

tectonic plate – a huge piece of rock that forms part of Earth's crust.

volcanic eruption – a bursting out of lava from a volcano.

INDEX

ONLINE RESOURCES

popbooksonline.com

Scan this code* and others like it while you read, or visit the website below to make this book pop!

popbooksonline.com/tsunamis

*Scanning QR codes requires a web-enabled smart device with a QR code reader app and a camera.